About this book

Some liquids are thin, others are thick. Some are heavy, some light, some form solids and others cause chemical reactions. You can make all liquids behave in magical ways by doing the experiments in this book. If you are not familiar with something you need for an experiment, look on pp.6-7 for an explanation. Always read 'Laboratory procedure' on p.7 before you start an experiment.

Contents

Cover illustration Tom Stimpson

Series editor Wendy Boase
Designer Amelia Edwards

First published 1982 by
Methuen Children's Books Ltd,
11 New Fetter Lane, London EC4P 4EE
in association with
Walker Books, 17-19 Hanway House,
Hanway Place, London W1P 9DL

© 1982 Walker Books Ltd
First printed 1982
Printed and bound by
L.E.G.O., Vicenza, Italy

Filmsetting by Crawley Composition Ltd

British Library Cataloguing in Publication Data
Watson, Philip
   Liquid magic. - (Science club; 1)
   1. Liquids - Juvenile literature
   I. Title   II. Boase, Wendy
   III. Series
   530.4'2   QC145.2

ISBN 0-416-24230-8

Liquid Magic

Written by
Philip Watson

Step-by-step illustrations by
Elizabeth Wood

Feature illustrations by
Ronald Fenton

METHUEN / WALKER BOOKS
London · Sydney · Auckland

Supplies and skills

Basic materials

You don't need a lot of special equipment to experiment at home. Clear a special shelf or cupboard and start collecting some basic, inexpensive materials for your home laboratory.

Clean, glass jam and pickle jars, pill, medicine and ink bottles.

Sheets of newspaper, brown paper or old wrapping paper.

Measuring jug, scales and set of cook's measuring spoons.

Old cups, saucepans, saucers, plates, glass bowls and jugs.

Old sieve and an eye-dropper.

Lolly sticks and wooden spoons.

Old table knife, sharp knife, penknife and scissors.

Ruler, pencils, cotton thread, string, paper clips and drawing pins.

Other materials

If you are not familiar with a particular material or piece of equipment you need for an experiment, look for it in the following list.

Balsa wood is very lightweight. It can be bought at a craft shop.

Borax is a water softener. Buy it at a chemist's shop.

Camphor has a strong, medicinal smell. Buy it at a chemist's shop.

Card and cartridge paper vary in weight. Buy these at an art shop.

Copper sulphate crystals can be bought from a chemist or the science section of a large toy shop. They are poisonous, so always wash your hands after touching them.

Dried milk, dried peas, dried yeast are foods. But them at a grocery or health food shop or a supermarket.

Epsom salts is a medicine. Buy it at a chemist's shop.

Filter funnels can be bought from a chemist, the science section of a large toy shop or a shop that sells wine-making equipment.

Food dyes colour food. Buy them at a supermarket or grocery shop.

Gelatine is a setting agent. Buy it at a supermarket or grocery shop.

Glass can be bought cut to size at a hardware shop or a shop that sells glass and mirror.

Glycerine is used in cake icing. Buy it at a supermarket or chemist.

Hydrogen peroxide is a bleach. Buy 20 volume strength at a chemist. Always wash your hands after touching it.

Lead weights can be bought from a shop that sells fishing tackle.

Liquid paraffin is a medicine. Buy it at a chemist's shop.

Masking tape won't leave a mark if peeled off gently. Buy it at an art or craft shop.

A metal punch can be bought at a hardware shop.

Methylated spirit is used to remove varnish. Buy it at a hardware shop.

Milk of magnesia is a medicine for upset stomachs. Buy it at a chemist.

A modelling knife has a very sharp blade. Buy it at an art or craft shop.

Mothballs ward off moths. Buy them at a supermarket or chemist's shop.

Muslin and cheese-cloth are gauzy fabrics sold at a haberdasher's or department store.

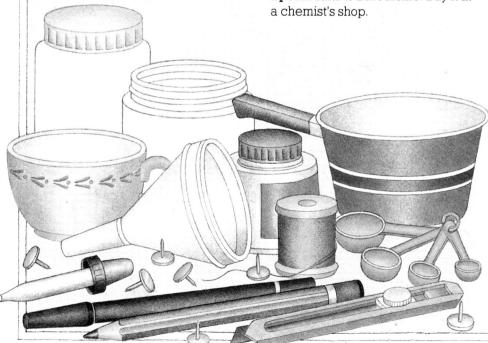

Oil-based printing inks are the kind that won't dissolve in water. Buy them at an art or craft shop.

Pliers or tin snips are sold at hardware or electrical shops.

Potash alum crystals can be bought from a chemist or the science section of a large toy shop. Always wash your hands after touching them.

Rennet is used to make cheese. Buy it at a health food or chemist's shop or at a delicatessen.

Sodium bicarbonate is used in cake-making. Buy it at a grocery shop or supermarket.

Sodium thiosulphate crystals can be bought from a chemist or the science section of a large toy shop. You should wash your hands after touching them.

Spirit-based felt-tipped pens have ink that will not dissolve in water. Buy them at an art or craft shop.

Tapers and spills are used for testing gases. Tapers are made of wax and spills of wood. Buy them from a tobacconist, a chemist or the science section of a large toy shop.

Test tubes can be bought from a chemist or the science section of a large toy shop.

Washing soda crystals are used to soften water. Buy them at a grocery shop or supermarket.

Water-based felt-tipped pens or inks are the kind that dissolve in water. Buy them at an art or craft shop.

White spirit dissolves gloss house paint. Buy it at a hardware shop.

Laboratory procedure

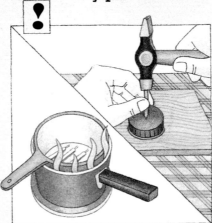

1. The exclamation symbol means that a tool (such as a metal punch), a material (such as hot wax) or a process (such as heating a liquid) can be dangerous. If you see this symbol on any part of an experiment, always ask an adult to read through the experiment with you before you start.

2. Put on old clothes, an overall or an apron before starting.

3. Read through an experiment, then collect the materials listed.

4. Clear a work area and cover it with newspaper or other paper. Put an old wooden chopping board or cork tile on the work area if you have to cut anything.

5. Take care not to get anything in or near your eyes. If this happens, immediately rinse your eyes in clean water, and tell an adult.

6. Never eat or drink anything unless told you may do so in an experiment.

7. Clean up any mess you make.

8. Wash your hands if you have touched a chemical, and when you have finished an experiment.

Drawing a circle

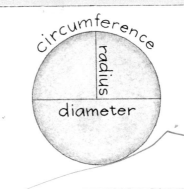

The radius is the distance between the centre of the circle and the circumference.
The diameter is double the radius.

1. Set the points of a pencil and a pair of compasses to the radius you require.

2. Put the point of the compasses firmly on a sheet of paper. It will mark the centre of the circle.

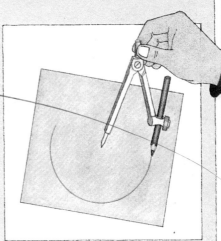

3. Swing the pencil round so that it draws a complete circle.

8

Moving liquids

Rain falls from the sky and rivers flow downhill – you never see rain or rivers moving upward. That's because an invisible force called gravity pulls everything towards the earth. Yet the experiments in this chapter show that liquids *can* flow against the force of gravity when certain scientific processes are at work.

Striped celery

One of the most important jobs of plant stems is to carry water from the roots up to the leaves and flowers. You can watch plants doing this in your home laboratory.

Materials

- 2 celery sticks, one with leaves, one without
- table knife
- glass of water
- blue or red food dye
- eye-dropper

1. Using the eye-dropper, add five drops of dye to the water.

2. Use the table knife to trim the bottom of each celery stick.

3. Put both celery sticks into the glass of coloured water.

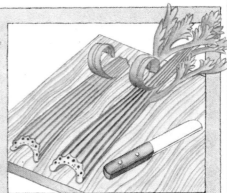

4. After about 30 minutes, peel away the outer skin of each piece of celery to see in which stick the colour has climbed higher.

The stems of the celery sticks are full of tiny hollow tubes. The coloured water climbs up these tubes by a process called capillary action. Trees also deliver water to their leaves by capillary action. Leaves themselves help suck up liquid. Their pulling power is really remarkable, as some branches may be 100 m from the ground.

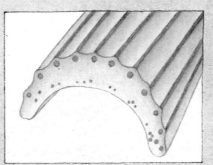

5. Cut across a celery stick to see the tubes.

6. Eat the celery, if you like.

Flower power

This experiment shows how a white flower changes colour by capillary action, or drawing up water through tiny tubes in its stem.

Materials
- fresh white carnation or dahlia with fairly short stem
- sharp knife
- 2 glasses water
- blue and red food dyes
- eye-dropper

1. Split the bottom half of the flower stem in two.

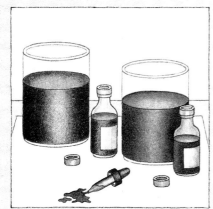

2. Using the eye-dropper, add 10 drops of red dye to one glass of water and 10 drops of blue dye to the other.

3. Put one part of the flower stem into the red water, and the other into the blue.

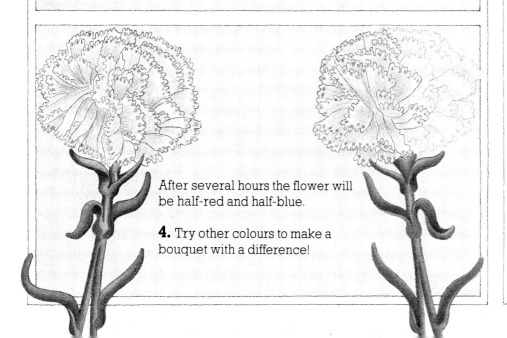

After several hours the flower will be half-red and half-blue.

4. Try other colours to make a bouquet with a difference!

Leafy ladders

Liquids other than water can rise up through the tubes in a plant's stem. Test this, and preserve some leaves at the same time.

Materials
- bunch of fresh, young beech or eucalyptus leaves
- vase of water
- glass jar and wooden spoon
- about ½ cup glycerine
- about 1 cup hot water
- table knife
- string

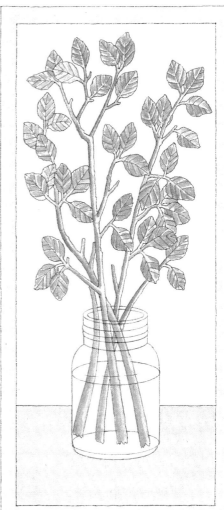

4. Put the stems into the glycerine and water mixture.

1. Put the stems into the vase of water and leave them overnight.

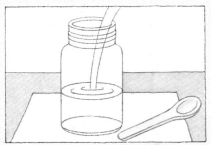

2. Put the glycerine and water into the jar, then stir the mixture. Allow it to cool.

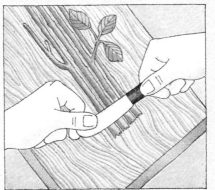

3. Take the leaves out of the vase and crush the bottom 1cm of the stems with the blade of the knife.

5. Put the jar aside and wait for the leaves to change colour. This may take a few days or weeks.

The glycerine and water mixture rises up the stems by capillary action. When it reaches the leaves, the water escapes into the air. But the glycerine, being oily, stays in the leaves, softening them.

6. Take the leaves out of the jar.

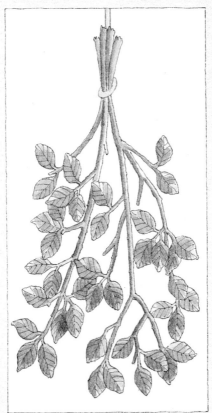

7. Tie the stems together and hang the leaves upside down for a few days. This helps the glycerine travel to the very tips of the leaves.

8. Arrange your preserved leaves in an empty vase.

Watery potato
Most foods contain water. You can see this if you cut a hole in a potato and put sugar in it. The sugar will dissolve as water comes out of the potato.

Bouncing peas
Some liquids can pass through the skins of certain foods. This process is called osmosis. When water moves through the skins of dried peas, something amusing happens.

Materials
- 250 g dried peas
- glass
- water
- tin plate

1. Fill the glass with dried peas, then top it up with water.

2. Put the glass on a tin plate.

The peas absorb the water through tiny holes in their skins. This makes them swell up, until they are too big for the glass. When they roll out of the glass, listen to the noise they make! (Keep topping up the glass with water to make all the peas roll out.)

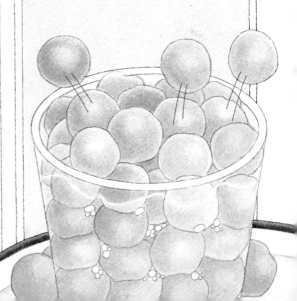

Spreading inks

This experiment shows that some ink is made of different-coloured chemicals which move at different speeds.

Materials
- white blotting paper 12cm square
- pencil, scissors and ruler
- water-based black, blue or blue-black fountain pen ink
- fine paint brush
- water and glass

1. Put the glass upside down on the blotting paper and draw round its rim with the pencil.

2. Cut out the circle, allowing about 1cm extra all round.

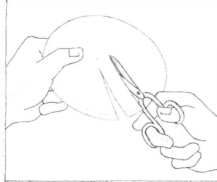

3. Make two parallel cuts from the edge of the paper to the centre, to form a strip about 1.5cm wide.

4. Put a large dot of ink on the middle of the paper and let it dry.

5. Fold the strip downward.

6. Almost fill the glass with water.

Rising inks

Colours in inks travel at different speeds in different liquids. You can test this, and make some lovely patterns at the same time.

Materials
- strips of white blotting paper about 0.5cm × 15cm
- water-based felt-tipped pens in dark colours
- glass of water
- glass jar of methylated spirit
- masking tape and a few books

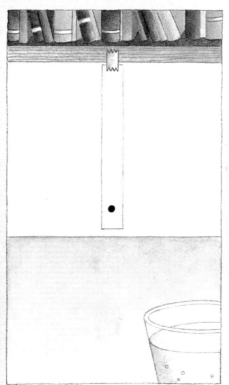

2. Tape the other end to a bench or cupboard so that the strip of paper hangs freely above a flat surface.

3. Put the glass of water below the strip of paper.

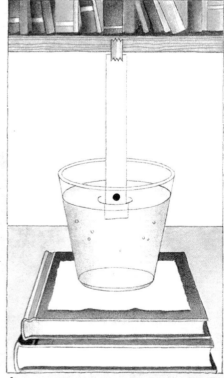

4. Pile books under the glass until the dot on the end of the strip is just above the surface of the water. Leave the paper in the glass for about 10 minutes.

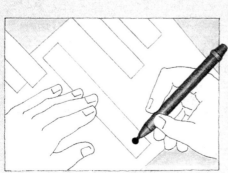

1. Put a dot of colour about 1cm from one end of a strip of paper.

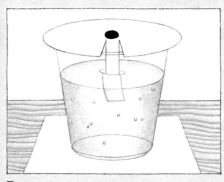

7. Rest the circle of paper on the rim of the glass, with the end of the strip in the water.
Leave it for about 10 minutes, or until the ink spreads to the edge of the paper.

The ink will dissolve in the water and split up into its different-coloured chemicals. Scientists call this chromatography. The faster-moving chemicals will move right to the edge of the paper. The slower ones will be left behind as rings round the central black dot.

8. Remove the paper from the glass and leave it to dry.

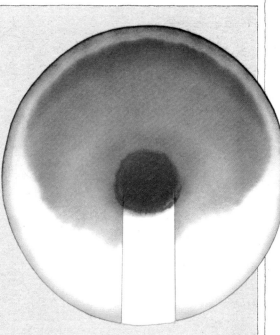

The different-coloured inks in the felt-tipped pen will dissolve in the water and spread up the blotting paper. The faster-moving ones will travel right to the top of the strip.

5. Remove the paper and let it dry.

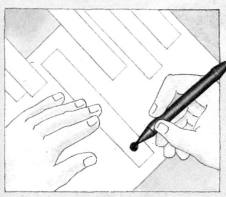

6. On another strip of paper, put a dark blue dot on top of a dark brown dot. (Or use other colours.)

7. Test this strip in the glass of water, as shown in step 4, to get a different pattern.

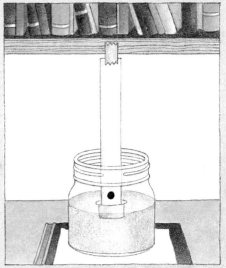

8. Repeat step 1, then suspend the paper in the jar of methylated spirit to make yet another pattern.

You could test different colours, as well as combinations of colours, using methylated spirit.

Use your strips of paper as book marks, if you like.

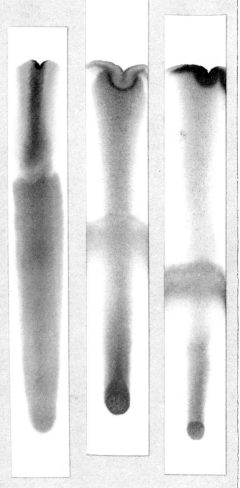

Speedy liquids

Liquids expand or contract in size according to the temperature. This is why thermometers contain liquids. Put a household thermometer into a glass of warm water, then into one of cold water. Watch how quickly the liquid moves as it records the different temperatures. Try taking the temperatures of other liquids.

Water thermometer

In this home-made thermometer, expansion and contraction of air causes water to rise or fall.

Materials
- metal punch and hammer
- small, glass bottle (such as ink bottle) with metal screw-top lid
- clear plastic straw
- plasticine
- small jug of water coloured with 4 or 5 drops food dye

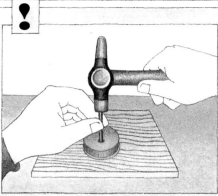

1. Ask an adult to punch a hole in the lid of the bottle. The straw should be able to fit into the hole.

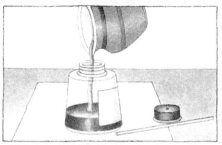

2. Pour the coloured water into the bottle until it is three-quarters full.

3. Screw on the lid.

4. Push the straw through the hole in the lid until about 1cm of its length is under the water.

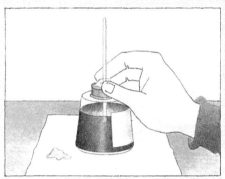

5. Use plasticine to seal the hole round the straw. This will stop air escaping from the bottle.

6. Cup both hands right round the bottle and hold it for several minutes.

As your hands warm the air in the bottle, it expands. This forces the water up the straw.

7. Take the lid off the bottle.

8. Tip water out until the bottle is only half-full.

9. Screw the lid back on and repeat step 6.

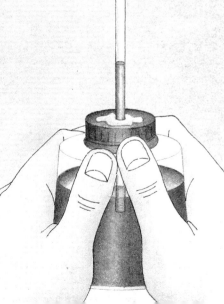

You will notice a difference. The water can be moved further if there is more air in the bottle.

10. Ask a few friends to hold the bottle. The hottest hands will make the water rise highest.

11. You could leave the bottle in different rooms to compare their temperatures.

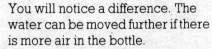

Milky volcano

In this experiment, a mixture of very ordinary liquids produces an effect rather like an upside-down volcano hurling out smoke.

Materials
● ice cube tray and milk
● glass cup of black tea or water

1. Pour milk into one section of the ice cube tray.

2. Put the tray into the freezer compartment of the refrigerator and leave it there overnight.

3. The next day, tip the frozen milk cube into the cup of tea (or water).

As the milk cube melts, you will see clouds of white liquid sink to the bottom of the cup. Milk is mostly made of water, and when iced water melts, it immediately becomes heavier.

Ice lollies

Some liquids – water, for instance – expand when frozen. See what happens by making ice lollies.

Materials
● small paper drinking cup
● stiff card about 10cm square
● pencil, scissors and ruler
● lolly stick
● tinned fruit juice and water

1. Put the drinking cup upside down on the card.

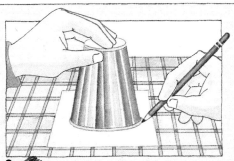

2. Draw round the rim of the cup with the pencil.

3. Cut out the circle, allowing about 1cm extra all round.

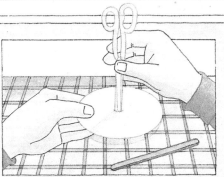

4. Use the tip of the scissors to punch a hole in the centre of the circle. The lolly stick should be able to fit into the hole.

5. Half-fill the cup with fruit juice.

6. Add water until the liquid rises almost, but not quite, to the rim of the cup.

7. Stir the mixture with the lolly stick.

8. Push the lolly stick through the hole in the cardboard circle.

9. Put the circular cover on the cup and push the stick down until it just touches the bottom of the cup.

10. Put the cup into the freezer compartment of the refrigerator. Leave it there overnight.

When you lift off the cover in the morning, you will see that the frozen fruit juice and water has expanded and completely filled the cup.

11. Tear away the paper cup and eat your ice lolly.

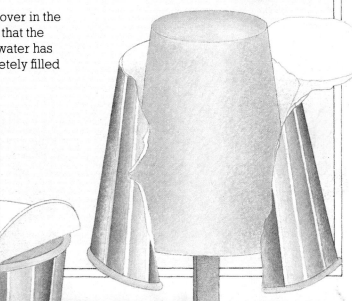

16

Floating liquids

Liquids, like people and objects, vary in 'weight'. Scientifically, they are said to have different densities. You have probably seen oil floating on puddles of water. That's because oil is lighter, or less dense, than water. In this chapter, you can find out the densities of many liquids, and see which ones will sink and which will float on others.

Liquid layers

Perfect bands of colour prove that some liquids are heavier than others.

Materials
- 3 sheets cartridge paper, each 30cm square
- pair of compasses
- pencil and ruler
- scissors and sticky tape
- tall, glass jar
- syrup or treacle
- ¼ cup water coloured with 4 or 5 drops food dye
- ¼ cup liquid paraffin
- ¼ cup methylated spirit
- small piece of candle, a marble and mothball

6. Carefully pour a 10cm-deep layer of syrup into the jar. It must not touch the side of the jar.

7. Hold one paper funnel in the jar at an angle, so that the pointed end rests against the side of the jar.

1. Using the compasses and pencil, draw a circle with a 15cm radius (see p.7) on each sheet of paper.

2. Cut out the paper circles.

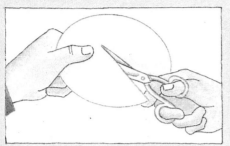

3. Cut half-way across each circle to the centre.

4. To make long, narrow funnels, fold each piece of paper round and round, then fix the outside edge with sticky tape.

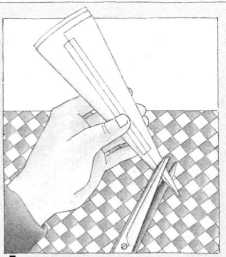

5. Snip off the point of each funnel to make a medium-sized hole.

8. Carefully pour in the coloured water, letting it drip through the funnel and down the side of the jar.

9. Use a second funnel and add the paraffin in the same careful way.

10. Slowly add the methylated spirit through the third funnel.

If you have added all the liquids very carefully, they won't mix. They will float in separate bands because they have different densities. Each of the liquids you have added is lighter, or less dense, than the one below it.

11. Gently lower the piece of candle on to the liquids in the jar.

The candle will float on top of the layer of coloured water. This means that the candle is heavier than paraffin and methylated spirit, but lighter than water and syrup.

12. Lower the marble and the mothball into the jar, to compare their densities with those of the liquids.

Solid tea joke

Make a cup of black tea and sprinkle two teaspoons of gelatine into it. Stir the mixture, then leave it to set overnight. At breakfast, offer it to someone who has a good sense of humour. The tea will look ordinary, but milk will simply float on top of it!

Coffee break

Next time you make coffee, let it cool a little and pour it into a clear glass. Slowly add a little milk that has been in the refrigerator overnight. Milk is denser than coffee, so it will sink. Try adding whipped cream to the coffee. The cream will float because air has been beaten into it, making it very light.

Hydrometer

Scientists use an instrument called a hydrometer to measure the density or 'heaviness' of liquids. Here's how to make a simple one.

Materials
- masking tape and scissors
- cigar tube or test tube
- small lead weights (the number depends on their heaviness)
- large jug of water
- clean cloth
- ruler and spirit-based felt-tipped pen or ballpoint pen

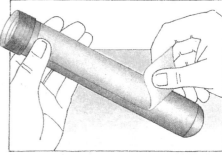

1. Cut a strip of masking tape and stick it on to the tube as shown.

2. Drop a few weights into the tube.

3. Put the tube into the jug of water.

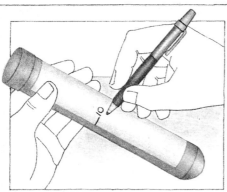

7. Write the number 10 against the water level mark. This indicates the density of water.

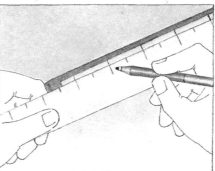

8. Using the ruler and pen, mark points 1cm apart up and down the tube from the number 10.

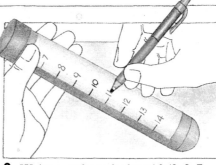

9. Write numbers below 10 (9, 8, 7, and so on) against the marks *above* the 10 mark. Put numbers above 10 (11, 12, 13 and so on) against the marks *below* the 10 mark.

Riding the waves

In this experiment, you can make a model boat ride to and fro on very realistic waves.

Materials
- large, screw-top glass jar
- jug of water coloured with 10 drops blue food dye
- about 350 ml liquid paraffin
- small piece of candle
- penknife.

1. Half-fill the jar with the water.

2. Cut out a simple model boat from the candle. (Or use the piece whole.)

3. Drop the boat into the jar.

4. Add liquid paraffin until the jar is as full as possible.

5. Screw on the lid.

6. Tilt the jar from side to side.

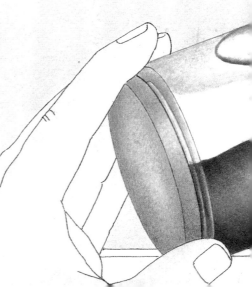

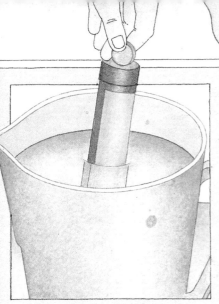

4. Add enough weights to make the tube float in an upright position.

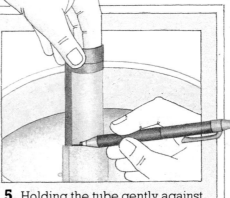

5. Holding the tube gently against the side of the jug, mark the water level on the masking tape.

6. Take the tube out of the water and dry it with the cloth.

10. Put the hydrometer into various liquids to compare their densities with that of water. You could test methylated spirit, liquid paraffin, vinegar, salty water or milk
Don't put the hydrometer into any liquid which someone might drink later on. It might be dirty.

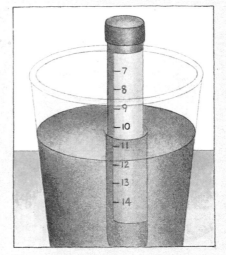

The boat will sail through the blue waves as they lap from one end of the jar to the other. You will notice that the waves roll very gently. That's because paraffin is thicker and lighter than water, and so slows down the water's movement.

7. To make more waves, simply tilt the jar from side to side again.

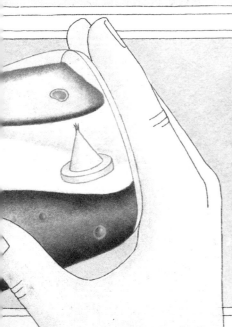

Swimming egg
This is a trick that will really astonish your friends.

Materials
- 2 glasses, one at least twice as large as the other
- water, salt and wooden spoon
- fresh egg

1. Half-fill the large glass with water.

2. Carefully lower the egg into it.

The egg will sink because it is heavier than water.

3. Fill the small glass with water and stir in salt until no more will dissolve.

4. Slowly pour the salty water into the large glass, taking care not to disturb the egg.

The egg will begin to rise because it is more buoyant in salt water than in fresh. In fact, salt water is denser than fresh water.

5. Add more water to make the egg sink, or more salt to make it swim.

❗ Wax sculptures

Plunging hot wax into water
produces fantastic sculptures.
Ask an adult to help you do this,
as hot wax can burn you.

Materials
- 6 white birthday cake candles
- coloured wax crayons
- penknife
- large saucepan of boiling water
- small, old saucepan
- sheet of clean glass 12cm square
- bucket of cold water

1. Put three white candles into the
small saucepan.

2. Put the bucket of water near the
stove, ready for following step 5.

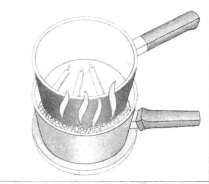

3. Lower the small saucepan into
the boiling water. Try not to get
water in the small saucepan or on
the wax.

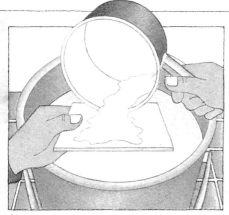

4. When the wax has melted, lift
out the wicks with the penknife,
then pour the wax on to the glass.

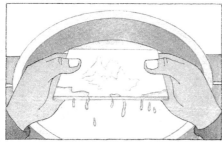

5. Holding the glass horizontally,
immediately plunge it into the
water. Lift the glass out again.

Wax, being lighter than water, tries
to rise to the surface as it cools. It
forms a sculpture as it does so.

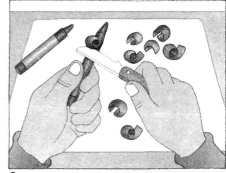

6. Shave some wax from one
crayon.

7. Put these shavings and the three
other white candles into the small
saucepan.

8. Repeat step 3. Jiggle the small
saucepan to mix the colour evenly.

9. Repeat step 4, pouring the hot
coloured wax over the white wax.

10. Repeat step 5.

11. Using white or coloured wax,
repeat the experiment as many
times as you like. The more layers
of wax you have, the more
spectacular your sculpture will be.

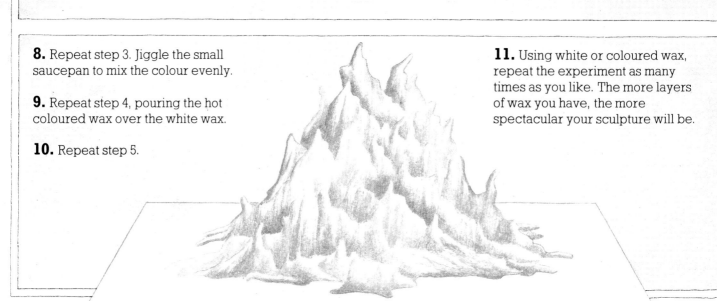

Marbling

The patterns in polished marble are often very beautiful. By using inks that float on water, you can imitate these patterns on paper. Look at the end sheets of this book to see some special effects.

Materials
- brown paper, old washing-up bowl and large jug of water
- white cartridge paper
- scissors
- 3 different-coloured oil-based printing inks
- 3 old saucers and 3 lolly sticks
- white spirit and teaspoon
- paint brush and metal comb

1. Cover the work area with brown paper.

2. Cut the cartridge paper into sheets that will fit inside the bowl. Put the sheets aside.

3. Pour cold water into the bowl to a depth of about 10cm.

4. Squeeze out about 3cm of one of the inks on to one saucer.

5. Add a teaspoonful of white spirit to the ink.

6. Stir the mixture with a lolly stick until the ink and spirit are well-blended.

7. Dip the paint brush into the mixture, then touch the surface of the water with the brush.

The ink should spread out. If it doesn't, mix a little more white spirit with the ink.

8. Wash the brush, then repeat steps 4-7 using the other two inks, saucers and lolly sticks.

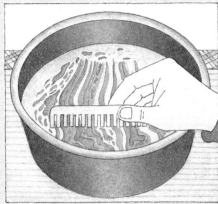

9. Now mix the colours by gently drawing the metal comb across the surface of the water.

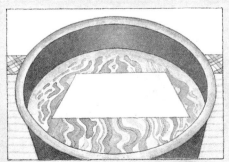

10. Gently lay a sheet of cartridge paper on the surface of the water.

11. Slowly lift off the paper.

12. Leave the marbled paper, pattern side up, to dry on the brown paper.

13. Brush more colour from each saucer on to the water in the bowl, and repeat steps 9-12. Or do step 14 as an alternative.

14. To get wavy patterns on both sides, pull the paper through the water, wiggling it as you do so.

22

Liquid mix

Heavy and light liquids – oil and water, for instance – often separate into layers. But such liquids will partly mix if an emulsifying agent is added. This is a substance which breaks down oil into tiny droplets so that it cannot form a layer. The mixture an emulsifying agent makes is called an emulsion. Experiment to see how emulsions are made and how they can be separated again.

Milky emulsion

Making an emulsion of oil and water is easy when you have the right emulsifying agent.

Materials
- screw-top glass jar
- 5 tablespoons cooking oil
- 5 tablespoons water
- washing-up liquid

1. Put the oil and the water into the jar.

2. Screw on the lid and vigorously shake the jar.

The two liquids will seem to mix.

3. Leave the jar for one minute and watch the liquids.

The liquids will slowly separate, and droplets of oil will begin to rise to the surface to form a layer.

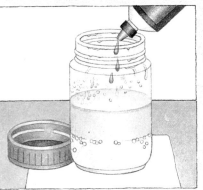

4. Now add about 10 drops of washing-up liquid. This is your emulsifying agent.

5. Shake the mixture until it begins to look rather like milk.

You have made an emulsion, in which the oil and water appear to be mixed perfectly. Under a powerful microscope, you would still be able to see tiny drops of oil in the water. But the washing-up liquid prevents the droplets of oil coming together to form a layer.

Greasy dishes

Try washing very greasy dishes in hot water, without using washing-up liquid. You will see that the grease floats on the water and the plates remain dirty. When you add washing-up liquid, the grease mixes with the water and the plates are cleaned.

Dressing a salad

Salad dressing is an emulsion of oil and vinegar. Next time you have a salad, make this tasty dressing.

Materials
- screw-top glass jar
- 1 tablespoon wine vinegar
- 5 tablespoons olive oil
- pinch of dry mustard
- salt and pepper

1. Put the oil and vinegar into the jar.

The two liquids won't mix.

2. Add a pinch of mustard.

3. Screw on the lid and shake the jar.

Mustard is an emulsifying agent, so it makes the oil and vinegar blend.

4. Add some salt and pepper and pour the dressing on your salad. Enjoy your meal!

Two-in-one snack

If you feel energetic enough to beat up an emulsion with a whisk, you can make something to eat and to drink at the same time.

Materials
- ¼ cup double cream
- small bowl and hand whisk
- small wooden spoon
- butter dish and clean glass
- salt

1. Pour the cream into the bowl and vigorously beat it with the whisk.

After a few minutes your arm may ache, but suddenly the cream will become solid.

2. Scrape the cream from the whisk and beat it a little longer with the wooden spoon.

Undrinkable milk

Milk is an emulsion of fat and water. You can see this if you split up the emulsion. Put a little milk into a small, glass jar. Leave it in a warm place until it turns very sour and separates into two layers. White lumps of fat will float on a pale, watery liquid that smells horrible!

Curd cheese

By splitting up an emulsion, you can make tasty cheese.

Materials
- 2 mixing bowls
- ½ litre lukewarm milk
- eye-dropper and 7 drops rennet
- wooden spoon
- muslin or cheese-cloth
- sieve
- salt

1. Pour the milk into one bowl.

2. Add the rennet and stir.

3. Cover the bowl with muslin and leave it in a warm place.

6. Tip the contents of the first bowl into the sieve to separate the curd from the whey.

7. Gather the edges of the muslin together and carefully lift it out of the sieve.

8. Squeeze the muslin over the bowl to press out most of the liquid in the curd.

3. Collect the solid on the spoon and put it into the butter dish.

Instead of an emulsion of fat drops in a liquid, you have an emulsion of liquid droplets in fat. This is butter.

4. Mix a little salt into the butter and it is ready for spreading.

The white liquid left in the bowl is buttermilk, which you can pour into the glass and drink.

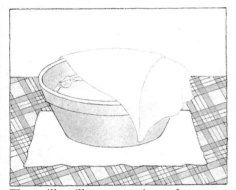

The milk will separate into a lumpy curd and a thin liquid called whey. (This may take 24 hours, or up to 48 hours.)

4. Line the sieve with the muslin.

5. Rest the sieve on the second bowl.

9. Put the muslin back into the sieve and leave the curd to drip overnight.

10. In the morning, add a pinch of salt to the curd and squeeze the muslin again.

Your curd cheese is now ready to be eaten. Try it with tomatoes, celery or cucumber. People used to eat the curd and whey together. (Taste the whey to see if you like it.)

Casein glue

The Ancient Egyptians probably used casein glue for building boats 4,000 years ago. This glue is made from milk, a common emulsion.

Materials
- 1 litre milk made from fat-free dried milk and water
- 1 cup vinegar
- saucepan and wooden spoon
- sieve lined with muslin or cheese-cloth
- newspaper and air-tight jar
- 20ml hot water in small jug
- 2 teaspoons borax

1. Pour the made-up milk and the vinegar into the saucepan.

2. Gently warm the mixture over very low heat for 10 to 15 minutes, or until a thick creamy-coloured scum forms on top of the liquid.

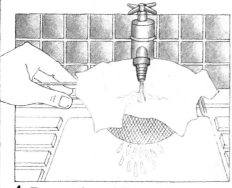

4. Turn on the cold tap and run water through the sieve to wash any vinegar off the curd.

5. Lift the muslin and gently squeeze out as much liquid as possible.

6. Put the curd between sheets of newspaper and leave it overnight.

7. Use your fingers to break up the lumps of dried curd into a powder.

8. Store the powdered curd in the air-tight jar.

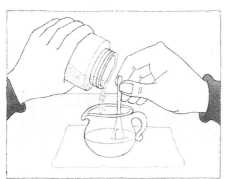

This liquid is casein glue, which can be used for many craft projects. It will stick sheets of paper together to make cards and collages, or pieces of light wood such as balsa, if the pieces are clamped firmly. The glue begins to harden after about an hour.

11. Wash the jug well and repeat steps 9 and 10 to make fresh glue.

10. Add about 25g of the curd and stir for three to five minutes, or until a white, creamy liquid forms.

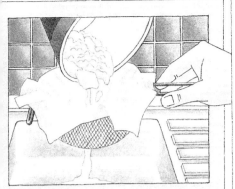

3. Hold the lined sieve over the sink and pour the contents of the saucepan into it.

Making yogurt

Bacteria are very simple forms of life which multiply rapidly if kept in a warm place. Because they are able to break down emulsions, they can turn milk into yogurt.

Materials
- 2 tablespoons live yogurt
- 1 cup milk
- saucepan
- small bowl and wooden spoon
- clean, dry cloth
- sugar and fruit

1. Pour the milk into the saucepan and gently heat it until it is lukewarm.

2. Pour the milk into the bowl.

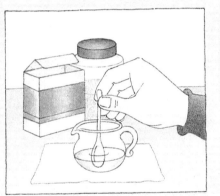

9. Vigorously stir the borax into the hot water for about five minutes. It should dissolve completely.

3. Add the yogurt to the milk and slowly stir the mixture until it is well-blended.

4. Cover the bowl with the cloth. Leave it in a warm place overnight.

Bacteria in the yogurt will multiply and curdle the milk. The next day, you will have a bowl of yogurt. (If not, leave the bowl longer.)

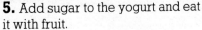

5. Add sugar to the yogurt and eat it with fruit.

Tough water

Solids have definite shapes, but liquids usually take the shape of things that contain them. If you put a few drops of water on a plate, they will spread out at first. But the water won't continue to spread. It will behave as if a 'skin' is holding it together, giving it shape. The surface of the water is said to have 'tension'. You can see surface tension at work by doing the experiments in this chapter.

Silver skier

This shiny little skier will float on water because surface tension is strong enough to bear its weight. But watch the skier move when you break the surface tension!

Materials
- aluminium foil 6cm × 12cm
- pencil and ruler
- scissors
- large bowl of water (or try this in your bath!)
- washing-up liquid

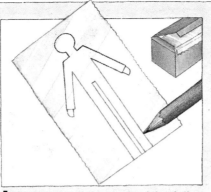

1. Using the full length of the foil, draw a figure roughly as shown. The legs should be about 8cm long.

2. Cut out the figure.

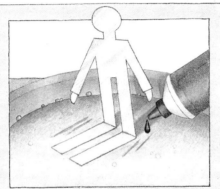

3. Fold the figure in half, then sit the top half up so that two 'skis' project in front of the body.

4. Gently lower the skier on to the surface of the water.

5. Put one drop of washing-up liquid behind the skier.

Washing-up liquid breaks the surface tension, and the skier shoots forward!

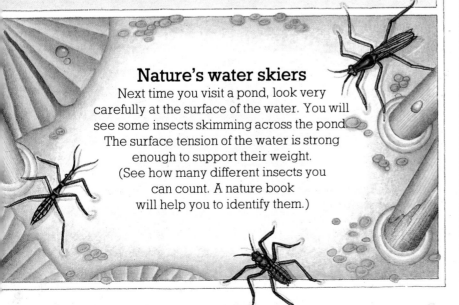

Nature's water skiers

Next time you visit a pond, look very carefully at the surface of the water. You will see some insects skimming across the pond. The surface tension of the water is strong enough to support their weight. (See how many different insects you can count. A nature book will help you to identify them.)

Water balls

Here's a way to make water roll round like a ball.

Materials
- 2 birthday cake candles
- large saucepan of boiling water
- small, old saucepan
- old tin plate
- penknife
- washing-up liquid

1. Put the candles into the small saucepan.

2. Lower the small saucepan into the boiling water. Try not to get water in the small saucepan or on the wax.

3. When the wax has melted, lift out the wicks with the penknife.

4. Pour the wax over the plate. Let it cool and solidify.

5. Hold the plate under a tap and let a drop of water fall on the wax.

The drop of water behaves as if it has a skin round it, and rolls across the plate like a ball.

6. Now drip some washing-up liquid on to the water.

The water ball will spread out as the washing-up liquid breaks the surface tension.

Blowing bubbles

When you blow bubbles, water really does look as if it has a skin. Blow your bubbles in sunlight, and you'll see pretty colours, too.

Materials
- 2 teaspoons washing-up liquid
- 1 cup water
- about 25cm fine, light wire
- plastic ruler

1. Add the washing-up liquid to the cup of water.

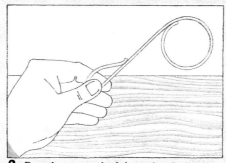

2. Bend one end of the wire into a loop and twist the other end into a handle.

3. Dip the wire loop into the cup and stir the liquid.

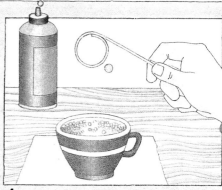

4. Carefully lift out the wire.

You will see a film of soapy liquid stretched across the loop.

5. Make a bubble by blowing gently through the loop.

6. Dip the loop into the liquid and blow again to get more bubbles.

If you have difficulty making bubbles, add more washing-up liquid.

You can even control your bubbles!

7. Rub the plastic ruler up and down a woolly sweater so that it becomes charged with static electricity.

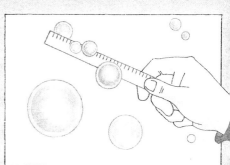

8. Blow a few bubbles and hold the ruler close to them.

Electrically charged particles in the ruler will attract the bubbles.

Camphor boat

This boat doesn't have a sail or an engine, yet it moves! Show a friend how to make one, so you can have boat races.

Materials
- balsa wood about 3cm × 5cm × 0.5cm
- modelling knife
- block of camphor about 1cm square

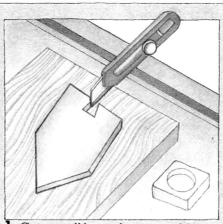

1. Cut a small boat-shape out of the wood, then cut a hole in the back as shown.

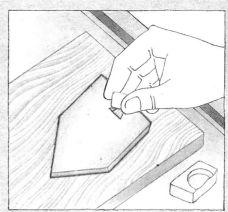

2. Cut a piece of camphor and plug the hole with it.

Moving match-sticks

This is a very magical trick. It works because soap and sugar affect the surface tension of water in very different ways.

Materials
- plastic or plastic-coated straw
- 1 saucer sugar
- 1 bar soft soap
- 4 dead match-sticks
- 2 bowls water

1. Wet one end of the straw and push it into the sugar.

2. Scrape the other end of the straw across the bar of soft soap.

3. Gently lay two match-sticks on the water in one bowl.

4. Dip the sugary end of the straw into the water between the two match-sticks.

Sugar absorbs water, so it draws the match-sticks together.

5. Repeat step 3, then step 4, using the soapy end of the straw.

Soap completely breaks the surface tension of water, so the match-sticks move apart rapidly.

Next time you take a bath, gently put the boat on the surface of the still water.

Your boat will begin to move slowly across the water. The camphor breaks the skin on the water, and so acts as a natural 'engine'.

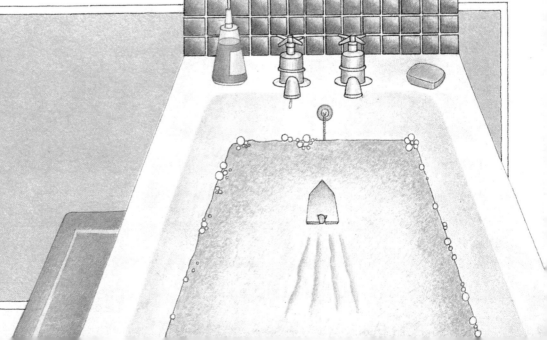

Liquid opposites

Many liquids contain chemicals which make them either acids or alkalis. Drinks which taste sharp, such as lemon or orange juice, are acidic. Liquids which feel soapy, such as washing-up liquid, are alkaline. (Liquids which are neither acidic nor alkaline – water, for instance – are said to be neutral.) The experiments in this chapter show how acids and alkalis behave and what tests can be used to distinguish them.

Fizzy lemonade

If two substances fizz when mixed, a chemical reaction is taking place. Here's a fizzy experiment that you can drink.

Materials
- lemon and sharp knife
- ½ litre water in saucepan
- sieve and jug
- clean glass
- 1 teaspoon sugar
- ½ teaspoon sodium bicarbonate

1. Slice the lemon, then put the slices into the saucepan of water.

2. Bring the water to the boil, and boil the lemon for about 10 minutes. Let the liquid cool.

3. Rest the sieve on the jug and pour the contents of the saucepan into it.

4. Remove the sieve and add the sugar to the lemon juice.

5. Pour some juice into the glass.

6. Stir in the sodium bicarbonate.

7. Drink the lemonade at once, before the bubbles escape.

The bubbles are carbon dioxide gas, which is given off when an acidic substance (lemon juice) reacts with an alkaline one (sodium bicarbonate).

8. Repeat steps 5-7 to make more lemonade.

Invisible ink

You can use lemon juice to send secret messages to your friends.

Materials
- lemon and sharp knife
- ½ litre water in saucepan
- sieve and jug
- fine paint brush
- sheet of writing paper
- iron and ironing cloth

1. To make the lemon juice, follow steps 1-3 of 'Fizzy lemonade'.

2. Remove the sieve.

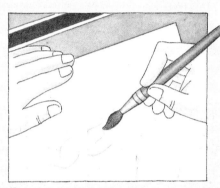

3. Dip the brush into the juice and paint a message on the paper. Leave the paper to dry.

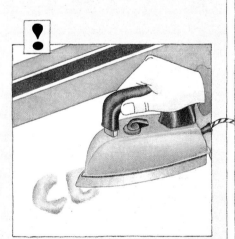

4. Run a hot iron over the paper.

Heat breaks down the juice into a brown chemical, which appears on the paper.

Dancing mothballs

Mothballs are quite heavy and would sink in most liquids. But in this fizzy combination of acidic and alkaline substances they will 'dance' in an amusing way.

Materials
- 5 or 6 small mothballs
- spirit-based felt-tipped pens
- large, glass jar and water
- 10 tablespoons wine vinegar
- 2 teaspoons sodium bicarbonate
- wooden spoon

1. Colour the mothballs with the felt-tipped pens.

2. Add water to the jar until it is three-quarters full.

3. Add the sodium bicarbonate and stir until it dissolves.

4. Add the vinegar and stir.

5. Drop the mothballs into the jar.

At first the mothballs will sink. Then they will dance upward. This happens because acidic vinegar and alkaline sodium bicarbonate react to make carbon dioxide gas. Bubbles of this gas collect on the mothballs. The gas, being lighter than water, lifts the mothballs to the surface. There the gas escapes, so the mothballs sink again, and the chemical reaction is repeated.

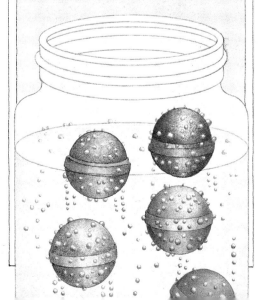

Indicators

Indicators are substances used to test liquids to see if they are acidic or alkaline. In this experiment, blackberry juice is used to make paper indicators for testing various liquids. A chart at the end of this experiment shows other juices you could use.

Materials
- tin of blackberries
- tin opener
- fine-mesh sieve and jug
- blotting paper about 10cm square
- scissors and spoon
- newspaper
- 2 screw-top glass bottles
- pencil and 2 sticky labels
- 1 teaspoon washing soda crystals dissolved in ¼ cup water
- eye-dropper
- about ¼ cup vinegar

1. Open the tin of blackberries.

6. Remove the strips with the spoon and put them on newspaper to dry.
Put the jug of juice aside.

7. Put all but one strip of the dyed papers into one bottle and label it 'alkali indicators'.

12. Put all but one of these strips into the second bottle and label it 'acid indicators'.

13. Drop the spare strip of paper into the vinegar.

The blue indicator paper will turn red. This shows that vinegar is an acid. All acids change acid indicator papers back to the original colour of the juice.

Test other household liquids to see whether they are acidic, alkaline or neutral. You could try salt dissolved in water, milk of magnesia or tomato juice, strong tea or milk. Don't drink any of the liquids you have tested.

2. Rest the sieve on the jug and pour the contents of the tin into it.

3. Remove the sieve and save the blackberries to eat later.

4. Cut the blotting paper into 20 small strips.
Put 10 strips aside.

5. Soak 10 strips in the blackberry juice until they turn deep red.

8. Drop the spare indicator paper into the washing soda solution.

The red indicator paper will turn blue, showing that washing soda is an alkali. These alkali indicator papers always turn blue in an alkali.

9. Try dropping an alkali indicator paper into the vinegar.

The paper won't change colour. This means that vinegar is either acidic or neutral. See which it is by making acid indicators from the reserved juice.

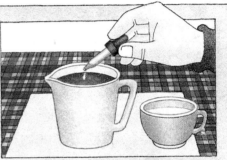

10. Using the eye-dropper, slowly add washing soda solution to the blackberry juice. Stop as soon as the juice turns dark blue.

11. Soak the 10 spare strips of blotting paper in this liquid, then dry them on newspaper.

Onion and red cabbage juices are also natural indicators. To make indicator papers from these, boil the vegetable in a small amount of water for about 10 minutes. Sieve the liquid and let it cool. Then follow steps 4-13. The chart shows how some natural indicators react.

Red violets
Violets are natural indicators. That is, they change colour in an acid. Put a violet into a glass or jar and pour some vinegar on it. The violet will turn red. This quick and pretty experiment shows that vinegar is an acid.

Indicator	Juice colour	Colour in alkali	Colour in acid
Blackberry	red	blue	red
Onion juice	colourless	green	colourless
Red cabbage	red	green	red

36

Liquids to gas

Enzymes are substances which cause chemical changes in living things. They control, for instance, the way in which the food you eat is changed into energy in your body. It is interesting to experiment with enzymes because they speed up many chemical reactions.

Liver enzyme

The enzyme in liver gives the fastest of all chemical reactions between an enzyme and another substance. Watch the reaction in this experiment.

Materials
- small, glass jar
- hydrogen peroxide
- 30-50g liver
- matches and taper or spill

1. Half-fill the jar with hydrogen peroxide. (Wash your hands.)

2. Drop in the liver.

Bubbles will immediately foam round the liver and rise up in the jar.

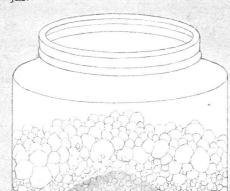

3. Light the taper, then blow it out to get a glowing end. Then hold the glowing taper in the neck of the jar.

The taper will relight with a soft pop, showing that oxygen is present. The oxygen is produced by the enzyme in the liver reacting with the hydrogen peroxide.

Yeast enzymes

In this experiment, yeast enzymes react with hydrogen peroxide to turn a liquid into a gas. (Always wash your hands after touching hydrogen peroxide.)

Materials
- small, glass bottle
- hydrogen peroxide
- ¼ teaspoon dried yeast
- shallow glass dish or bowl
- several coins
- small piece of card
- matches and taper or spill

1. Put the dish somewhere light and warm.

2. Stack the coins in it in two equal piles. The neck of the bottle should be able to rest on the coins.

3. Pour enough hydrogen peroxide into the dish to cover the coins by about 1cm.

4. Almost fill the bottle with hydrogen peroxide.

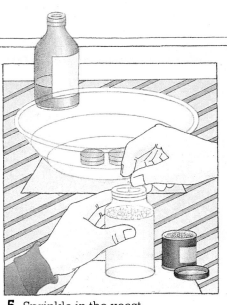

5. Sprinkle in the yeast.

6. Cover the bottle with your hand and tip the bottle a few times.

7. Hold the card over the bottle and gently turn it upside down.

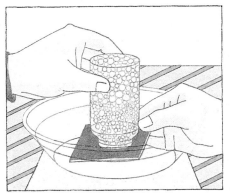

8. Lower the bottle into the dish so that its neck is below the surface of the hydrogen peroxide.

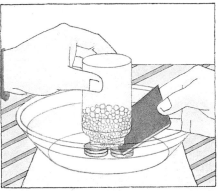

9. Slide away the card and rest the neck of the bottle on the coins. Leave it for about two hours.

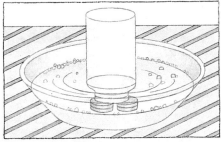

A gas will form in the bottle and slowly force out the liquid.

10. When the bottle is empty, slip the card under it again and lift it out. Don't let any gas escape.

! **11.** Light the taper, then blow it out to get a glowing end. Then slip away the card and plunge the taper into the bottle.

It will relight with a soft pop. The yeast enzymes have reacted with the hydrogen peroxide to form oxygen.

38

Fresh bread

Yeast is useful to cooks, as the enzyme in it makes bread expand and rise before it is baked. See what hard workers enzymes are!

Materials
- small cup, large mixing bowl and wooden spoon
- ½ teaspoon sugar
- 100 ml warm water
- 1 teaspoon dried yeast
- 200 g plain flour and a little extra flour for sprinkling
- pinch of salt
- large wooden board, large plastic bag and baking tray
- cooking oil
- 50g dried fruit or nuts (optional)

1. Put the sugar and warm water into the cup.

2. Stir in the yeast. Leave the mixture in a warm place for 15 minutes.

As the yeast enzymes go to work, a foam will form on top of the liquid.

6. Dip your hands in flour, then mix the ingredients together until they are no longer sticky. Add more water if the mixture is too dry, or more flour if it is too wet.

7. Sprinkle a little flour on the board and put the dough on it.

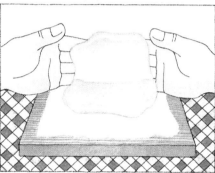

8. Fold the bread dough in half towards you.

9. Then push the dough away from you with your fists. Repeat steps 8 and 9 for about five minutes, turning the dough every few seconds. This 'kneading' makes dough thick and stretchy.

12. Set the oven temperature to 230°C (450°F or gas mark 8).

13. Take the dough out of the bag and shape it into small rolls.

14. Put the rolls on the baking tray, then put the tray into the oven for 20 or 30 minutes, or until the bread rolls are brown on top. Take the tray out of the oven and let the bread rolls cool.

To make fancy bread rolls, add dried apricots, currants or nuts to the flour before pouring in the yeast.

3. Put the flour and salt into the mixing bowl and make a hole (or 'well') in the flour with the spoon.

4. Pour the frothy yeast liquid into the well.

5. Stir it in with the spoon.

10. Lightly oil the plastic bag and the baking tray.

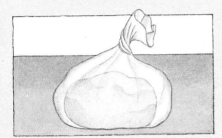

11. Put the dough into the bag and loosely twist the neck closed. Leave the bag in a warm place for about 45 minutes.

The dough will gradually increase its bulk until it has doubled in size. This happens because the yeast reacts with the sugar to make carbon dioxide, and this gas inflates the dough.

Holes in bread

When you cut a slice of bread from a loaf, you can see that it is full of tiny holes. These holes are made by carbon dioxide, the gas produced when yeast enzymes react with sugar.

Gas from fruit

Find out what kind of gas is made when yeast reacts with fruit juice. (This is how beer is made.)

Materials
- apple or orange peeled and sliced
- 1 cup water in saucepan
- sieve and jug
- small, glass bottle
- ½ teaspoon sugar
- ¼ teaspoon dried yeast
- taper, matches and balloon

1. Simmer the fruit in the water for five minutes.

2. Pour the liquid through the sieve into the jug and allow it to cool.

3. Remove the sieve and fill the bottle with liquid.

4. Add sugar and yeast.

5. Cover the bottle with your hand and shake the mixture.

6. Put the balloon on the bottle and leave it in a warm place overnight.

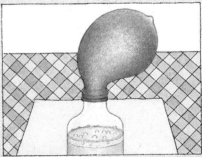

In the morning, the balloon will be partially filled with carbon dioxide, produced by the reaction between yeast enzymes and fruit juice.

7. Light the taper, pull off the balloon and plunge the taper into the bottle.

Carbon dioxide doesn't support burning, so the flame will go out.

Crystals

Many substances are made of crystals, which are particles arranged in regular, geometric patterns. Diamonds are composed of crystals of carbon. Salt and sugar are also composed of crystals. You can see the crystalline nature of salt or sugar if you look at a few grains of these substances through a microscope. Crystals are fascinating and beautiful, as you will find by doing the experiments in this chapter.

Bath-time crystals

All crystals contain water, even if they feel dry. Because washing soda crystals readily take up liquid, you can dye them and sprinkle them into your bath-water. Be sure to use the exact quantities given; too much colour will dye your skin.

Materials
- ½kg washing soda crystals
- large plastic bag
- eye-dropper and food dye
- few drops of water
- few drops of perfume (optional)
- large sheet of paper
- air-tight glass jar

1. Put the crystals into the bag.

2. Add the water and two drops only of dye.

3. Hold the neck of the bag closed, and shake it until the colour spreads through the crystals.

4. Add perfume for scented crystals, and shake the bag again.

5. Empty the crystals on to the paper and leave them for 10 minutes. Don't leave them any longer, or they may become powdery.

6. Put the crystals into the jar.

When you run a bath, sprinkle a handful of the coloured crystals into the water.

Crystals in nature

When a solid dissolves in a liquid, such as
salt in water, it disappears. If the liquid evaporates,
or dries up, the solid reappears. You can't see the salt
in sea water, but look for salt crystals in a rock pool
after the sun has dried up the water.
To get the same effect, put some salt into a shallow saucer
of water and leave it near a heater for two or three days.

Frosted glass

In this experiment, you can watch crystals form an unusual, needle-like pattern. The crystals seem to 'grow', but they are really crystallizing out of a liquid. All crystals are solids, but to make crystals you have to start with a liquid.

Materials
- 2 tablespoons water in saucepan
- wooden spoon
- about 5 tablespoons Epsom salts
- clear, soluble liquid glue
- sheet of glass 15cm square
- wad of cotton wool

1. Bring the water to near-boiling point, then lower the heat and stir in Epsom salts a little at a time. Keep adding salts and stirring until no more will dissolve.

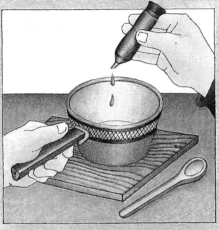

2. Remove the pan from the heat and let the liquid cool a little. Then add two drops of glue, stirring until it has dissolved.

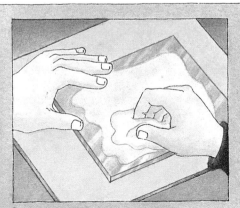

3. Dip the cotton wool into the mixture, then gently draw it across the sheet of glass so that the mixture is spread evenly.

Now see how quickly crystals can form. As the water in the mixture dries up, a pattern of needle-like crystals will cover the glass. These are crystals of magnesium sulphate, which is the chemical name of Epsom salts. (Hold the glass to the light to see the shapes clearly.)

White needles

Sodium thiosulphate, unlike most crystals, melts when heated. This means you get surprisingly rapid crystallization.

Materials

- small cardboard box and penknife
- about 30cm strong wire
- about 20g sodium thiosulphate crystals and test tube
- saucepan of boiling water

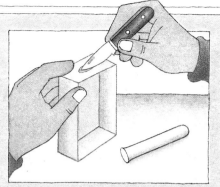

1. Cut a hole the size of the test tube in one side of the box.

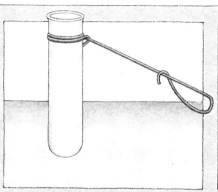

2. Twist one end of the wire twice round the neck of the test tube and loop the other end into a handle. Make sure the tube is held firmly.

3. Drop crystals into the test tube until they are about 3cm deep. (Wash your hands.)

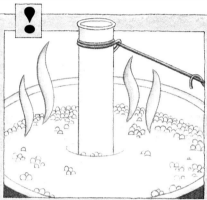

4. Holding the wire handle, lower the test tube into the boiling water. Don't let water into the tube.

5. When the crystals have melted, take out the tube and prop it in the hole in the box.
Leave it there for 15 minutes.

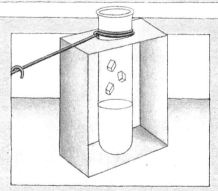

6. Drop three more of the sodium thiosulphate crystals into the tube. (Wash your hands.)

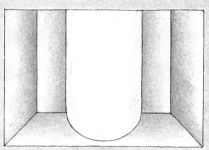

The liquid will rapidly solidify until the test tube contains a forest of needle-like crystals.

Blue giant

You can see very beautiful crystalline shapes if you grow a really large, brilliantly blue crystal from fine crystals of copper sulphate. Copper sulphate is poisonous, so do not get it near your eyes or mouth, and always wash your hands after touching it. Copper sulphate also damages metal, so make sure that the saucepan and knife you use for this experiment are old ones.

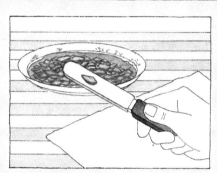

5. Gently lift out a small crystal with the table knife and put it aside on the sheet of paper.

11. Slowly pour the solution into the lined filter funnel. Don't let any undissolved crystals go with it.

12. Remove the funnel.

Materials

- 100 ml water in very old saucepan
- small, glass jar
- about 150 g copper sulphate
- wooden stick and old table knife
- shallow glass dish
- sheet of writing paper
- cotton thread, scissors and pencil
- filter funnel
- circle of blotting paper about 7cm in diameter

1. Boil the water and then pour it into the small, glass jar.

2. Add the copper sulphate, a little at a time. Stir and crush it with the stick until no more will dissolve.

3. Carefully pour the solution into the glass dish.

4. Put the dish in a warm place for one or two days, or until crystals about 0.5cm in diameter form.

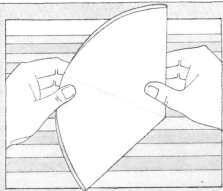

6. Fold the circle of blotting paper in half, then in half again.

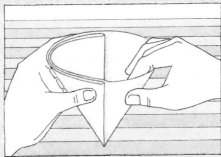

7. Put one hand round the folded paper. With your other hand, separate the folds to make a funnel.

8. Fit the blotting paper funnel into the filter funnel, then rest the filter funnel in the glass jar.

9. Tip the solution and any crystals in the dish into the saucepan.

10. Gently reheat this mixture, stirring until most (or all) of the crystals have dissolved.

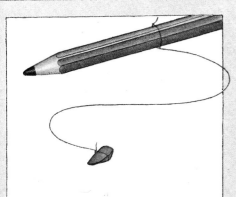

13. Tie one end of the thread round the crystal you have put aside, and trim the knot. Tie the other end to the pencil. (Wash your hands well.)

14. Balance the pencil on the jar so that the crystal hangs in the liquid. Leave the jar in a warm place.

If the crystal dissolves, or doesn't grow in a few days, the solution was not concentrated enough. Reheat, and dissolve more crystals in it.

15. When the crystal is as big as you want, pull it out of the jar.

You can reheat the solution, dissolve more crystals in it, and repeat the experiment if you like.

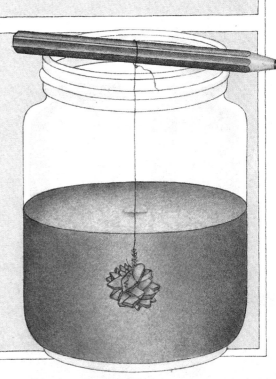

Sugar-candy

This is a delicious experiment because you can eat the result!

Materials
- 1 cup water in saucepan
- about 400 g white sugar
- wooden spoon
- thick, glass jar
- pencil or stick
- string and heavy button

1. Bring the water to boiling point, then lower the heat and stir in some sugar. Add more sugar and stir until no more will dissolve.

2. Let the liquid cool a little, then pour it into the jar.

3. Tie one end of the string to the pencil and the other to the button.

4. Balance the pencil on the jar. The button will keep the string hanging straight down in the liquid. Leave the jar in a warm place.

As the liquid cools, the sugar will crystallize on the string and button. In a few days, you can pull out the candy and eat it.

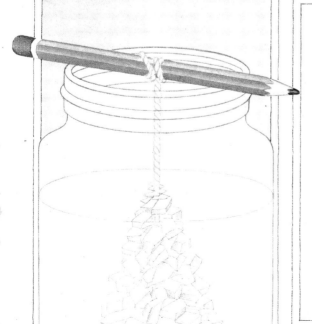

Crystal shapes

Crystals will form round any central point, or nucleus, so it is easy to create crystalline sculptures. Copper sulphate is poisonous, so do not get it near your eyes or mouth, and always wash your hands after touching it. If you use potash alum crystals, which are mentioned at the end of the experiment, handle them with the same care.

Materials
- 4 pipe cleaners
- cotton thread
- pair of pliers or tin snips
- 200 ml water in saucepan
- about 250 g copper sulphate
- wooden stick and pencil
- 2 × 200 ml glass jars, one with a wide neck
- filter funnel
- circle of blotting paper about 7cm in diameter

1. Cut the pipe cleaners in half.

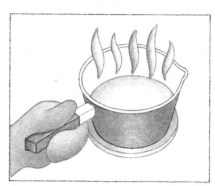

6. Heat the water in the saucepan to near-boiling point.

7. Allow it to cool a little, then pour it into the narrow-necked jar.

8. Add the copper sulphate, a little at a time. Stir and crush it with the stick until no more will dissolve.

12. Let the solution cool, then put the jar on a sunny window sill or in a warm place.

13. Tie the spare end of the thread attached to the tree to the pencil.

14. Balance the pencil on the jar so that the tree hangs in the solution. Leave the jar in a warm place for two or more days.

As the water evaporates, crystals will grow on the tree. If this does not happen, pull out the tree and reheat the solution, dissolving more crystals in it.

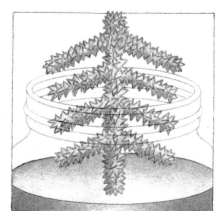

15. When your model is fully crystallized, carefully pull it from the jar. Treat the sculpture gently so that crystals do not break off.

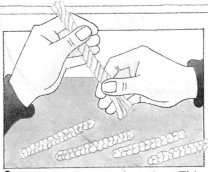

2. Twist four lengths together. This is the trunk of a model tree.

3. Tie one end of the cotton thread to one end of the tree trunk.

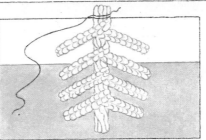

4. Twist each of the remaining short lengths round the trunk to make the tree's branches.

5. Clip the ends of the branches if they are uneven.

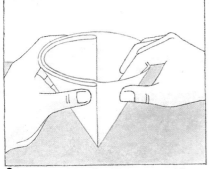

9. Use the blotting paper to make a funnel as shown in steps 6 and 7 of 'Blue giant' (see pp.42-43).

10. Fit it into the filter funnel.

11. Slowly pour the copper sulphate solution into the lined funnel. Don't let any undissolved crystals go with it, or more crystals will grow on these instead of on the model tree.

You can use the solution again to crystallize a different model by reheating and dissolving more copper sulphate in it. (You can also store the solution in a jar.)

There are many shapes you could use as models (see right), and many types of crystals you could use in a solution. Try Epsom salts (200 g dissolved in 100 ml hot water) or potash alum (250 g in 100 ml hot water). These crystals are colourless, but you can add four or five drops of food dye to the solution, to make lightly tinted pipe cleaner models.

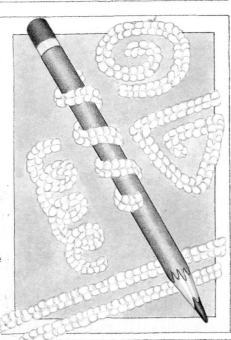

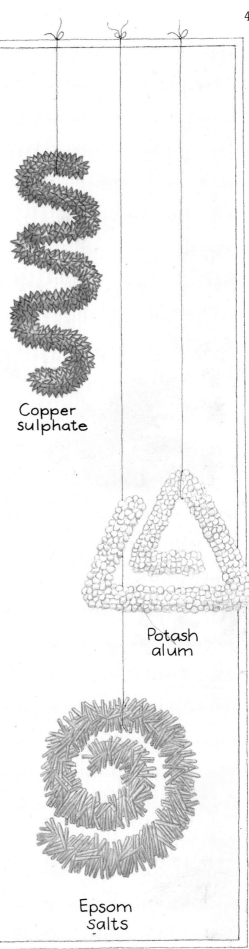

Copper sulphate

Potash alum

Epsom salts

Glossary and index

Words in CAPITAL LETTERS are also defined in the glossary.

acid
A sharp-tasting substance which turns INDICATORS a characteristic colour.　*see pp.32-33, 34-35*

alkali
A soapy-feeling substance which turns INDICATORS a characteristic colour.　*see pp.32-33, 34-35*

bacteria
Tiny forms of life which multiply rapidly if kept warm and fed.
　see p.27

capillary action
The process by which a liquid climbs up a hollow tube.
　see pp.8-9, 10-11

carbon dioxide
A gas which will extinguish a glowing taper. Its chemical formula is CO_2.
　see pp.32-33, 34, 38-39

chemical reaction
The process by which one or more substances react to form new substances.
　see pp.32-33, 34, 36-37, 38-39

chromatography
The process by which a substance dissolves and splits up into its component chemicals.
　see pp.12-13

crystal
A solid substance made of particles arranged in a regular, geometric pattern.　*see pp.40-41, 42-43, 44-45*

curd
A clotted substance formed by the CHEMICAL REACTION between an ACID and milk.　*see pp.24-25, 26-27*

density
The mass of a substance in relation to its volume (the amount of space it takes up).
　see pp.16-17, 18-19, 20-21

emulsion
A mixture of two liquids in which one liquid exists as tiny droplets suspended in the other liquid.
　see pp.22-23, 24, 25, 26-27

enzyme
A chemical found in living things which speeds up a CHEMICAL REACTION.　*see pp.36-37, 38-39*

hydrometer
An instrument used for measuring the DENSITY of a liquid.
　see pp.18-19

indicator
A substance which changes colour in an ACID and in an ALKALI.
　see pp.34-35

neutral
A substance which is neither an ACID nor an ALKALI.　*see pp.32, 34*

osmosis
The process by which liquids pass through certain membranes, or skins.　*see p.11*

oxygen
A gas which will relight a glowing taper. Its chemical formula is O_2.
　see pp.36-37

surface tension
A force causing the surface of a liquid to behave as if it has an elastic skin on it.　*see pp.28-29, 30-31*

whey
A pale yellow liquid formed when milk curdles.　*see pp.24-25*